# Forest and Garden

LIVING
NATURE
SERIES

## William H.
## White Jr., Ph.D.

PHOTOGRAPHS BY THE AUTHOR

 STERLING PUBLISHING CO., INC. NEW YORK

Oak Tree Press Co., Ltd. London & Sydney

## OTHER BOOKS BY THE SAME AUTHOR

## OTHER NATURE BOOKS

The author and publisher wish to thank James M. White and Elizabeth J. White for their many hours of searching in the woods and tending the garden; William White, III and Sara Ann White for conducting the experiments necessary for this book; Margaret R. White for her untiring efforts in the pursuit of birds and insects; and Rebecca L. White for preparing the manuscript.

Library of Congress Catalog Card No.: 76–19799
Sterling ISBN 0-8069-3578-2 Trade      Oak Tree 7061-2516–9
3579–0 Library

# Contents

Color section follows page 16

Illus. 1. The leaves of the tall trees absorb the light and heat of the midsummer sun, while the lower levels of the forest stay dark, cool and moist.

# Introduction

The temperate zone of the northern hemisphere covers vast areas and is the habitat of the great deciduous or "leaffall" forests. In fact, from remote prehistoric times, the natural state of most of the land surface of the region has been forest. Remnants of that forest still cover vast areas of the United States, Canada, Scandinavia, Russia, China and Japan. Vast stretches of the forest of the north temperate zone have been cleared for farms and human habitations, for towns, pastures, orchards and gardens. One of the most interesting aspects of the zone is the contrast between the forest and the cultivated land, or garden, which adjoin one another over long stretches. The forest and the garden consist both of horizontal communities (similar organisms of different species) and vertical communities (dissimilar organisms of different genera). The communities of plants and animals are interdependent and interrelated.

This book describes the natural processes that create and maintain the forest and the artificial techniques by which man creates and maintains the garden, and examines key life forms to be found in each.

Illus. 2. Soon after sunrise, the morning light illuminates the undergrowth along a forest trail.

Illus. 3. The forest floor is littered with a deep layer of humus being turned into fertile soil by the digestion of bacteria and the ploughing action of earthworms.

# The Forest

A forest is a mighty living thing. It is the most durable and final of living communities. It is so dynamic and tenacious of life that it can alter the face of the land and even change the climate. Although a forest may be burned over or all of the trees may be cut down, it can still restore itself. It can grow up again to resemble very closely its original character. A forest can spread and conquer new areas through the slow but invincible advance of plant *succession*—the replacement of one type of vegetation by another, usually over a long period of time. A forest can resist and survive cold, heat, wind, drought, flood and fire. It can withstand almost any natural force or catastrophe.

A forest is far more than trees. A forest is the whole community of life forms—living, breathing, germinating, growing, maturing, reproducing and dying. Animals such as squirrels and simple plants such as lichens are just as much components or parts of the forest as are the trees. A forest changes each hour of every day throughout the year. These changes are due to the amount and type of light available and to various factors arising from the

Illus. 4. The great branches of large trees hold up a massive surface area of leaves to the sun and serve as pipelines for water and dissolved minerals.

climate. Yet the forest can affect the climate, for it can enclose great volumes of air and capture large quantities of water and so slow down the more abrupt climatic changes.

On a hot summer day the forest appears dark and cool. When the roads and fields are dry and parched, the forest can be moist and even humid. Here and there, a gap in the canopy of leafy branches allows a dazzling shaft of sunlight to illuminate the darker green leaves of plants on the forest floor. At midafternoon when the sun is at its highest and the air is still and hot, all of the mammal and bird life of the forest ceases and only the insects are left to buzz about. In the early morning and the cool evening, the sun's rays shine obliquely and the light becomes redder with the sun's setting until the whole forest is a deep crimson just as nightfall descends. The cooler night air begins to draw up water vapor as mist from the forest floor. It is at this time that animals such as birds, reptiles and even the insects are at their peak of activity.

## Plants and Animals

The climbing mammals, birds and other creatures which live among the tallest trees are easy enough to see. Less visible are the vibrant and bustling communities which live in the humus on the forest floor. The humus, formed from layers of fallen leaves from previous seasons, is fairly humming with the activity of hundreds of species of micro-

Illus. 5. The highest branches of the trees float in air which is continually suffused with sunlight.

Illus. 6. Bright green crabapples ripen to a bright red in the late summer sun.

organisms, as well as species that are larger but still small enough to elude the casual observer.

The popular view of the distinctions between plants and animals holds that plants stay in one place, that they have roots in the soil and green leaves, and that they are not sensitive as animals are. However, there are some other differences which may be less obvious but even more important. All of the processes which animals carry on by mechanical means—moving, eating, fighting—plants must do by chemical means. For example, animals must move about searching for food and in the process they must pursue, catch, kill and devour it. An animal must prey on

Illus. 7. The three stages of a fallen hickory nut. On the left is the newly fallen nut still in its heavy green seed coat, on the right a nut from the last season's fall with its seed coat dried and split open, and in the middle the completely peeled kernel.

Illus. 8. At the edge of the woods, the grass from the meadow and the young trees and shrubs mingle and intermix environments.

Illus. 9. A game trail winds through a hedgerow under a tunnel of trees.

other living organisms to get the minerals its system requires, for it cannot absorb them directly itself.

On the other hand, a plant puts off enemies by emitting a bad smell, or having an unpleasant taste, or even secreting poison in its leaves or fruits by chemical means. The fact is that plants are extremely complicated and differ greatly in the number and type of compounds which they manufacture from very simple materials, usually little more than oxygen, carbon, nitrogen, sulphur, different metallic substances and water.

## The Trees

The most visible component of the forest, the trees, are actually gigantic food factories where tons of water,

Illus. 10. An open space beneath the larger trees is hidden from the roadside by the leaves of younger trees and shrubs.

Illus. 11. A forest has been cut back along the lawns of a public park.
The natural margin of younger trees and shrubs is destroyed by continual
grass cutting along the edge of the woods.

hundreds of pounds of minerals, and thousands of cubic
feet of gases are turned into foodstuffs and necessary chemicals, with energy derived from the sun, through the process known as *photosynthesis*. Photosynthesis is the process
in which the green coloring matter in leaves, called chlorophyll, aids in the manufacture of carbohydrates from
carbon dioxide and water. The roots of the largest tree
may extend wider and farther than any of the branches.
The trunks provide storage space for hundreds of pounds
of carbohydrates bound up in the form of *cellulose*, the

basic tissue of plants. They also hold up the branches into the air, so that the thousands of leaves can be exposed to the sunlight.

The great branches support and spread the canopy of leaves in order to expose the greatest possible area to the sun's radiation, for it is in the layer of cells just below the top surface of each leaf that the vital process of photosynthesis takes place. While the branches of the forest trees look as though they grow every which way, they actually grow at the precise angles and to the exact lengths required to achieve the most exposure. The leaves and branches of one tree will grow in precise adjustment to those of the trees around it so that the maximum sunlight is absorbed by all. In addition, the trunk and the branches must grow in diameter in proportion to the increase in length so that they can support the weight of the leaves. The highest branches are awash in a constant ocean of air.

It is the very topmost or *apical* buds of the topmost branches which control the tree's further growth. While animals have nervous systems which cause them to react, plants have systems of secreted chemicals called *hormones*. These chemicals carry messages to the cells of the tree or plant and cause the increase or decrease of activity to meet some need or react to an outside stimulation. It is the hormonal system which causes the buds to grow in the spring, forces the central topmost branches to grow fastest, and causes the leaves to fall from the branches in the autumn.

Illus. 12. The forest was cleared from this land and it was farmed for many years, but farming was halted and the secondary growth of trees is beginning to turn the area back into young forest again.

## Gymnosperms and Angiosperms

Around the base of each tree a whole unique environment exists. Shelter, shade and moisture are provided for specific communities of plants and animals. The deciduous trees produce their seeds in the autumn, which winter over at the base in the mulch of leaves and humus. Trees produce many types of seeds. Two of the main divisions of trees on the basis of seed type are the *gymnosperms* from the Greek term meaning "naked seeds" and the *angiosperms* from the Greek meaning "covered seeds." The gymnosperms have small, completely uncovered seeds often

Plate 1. A close-up photograph of a moss clump shows the fruiting bodies rising above the clump.

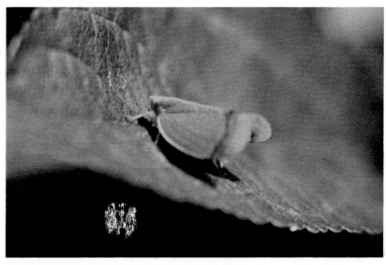

Plate 2. A green leaf-hopping insect nearly blends into the foliage upon which it feeds.

A

Plate 3. Looking like a bright wildflower, the cap of this bright yellow fungus grew up in a mere 24 hours.

Plate 4. A clump of wild Indian ginger *(Asarum canadense)* grows in a tiny forest clearing.

Plate 5. A yellow crab spider searches for prey.

Plate 6. Hatchling robins wait impatiently for food.

Plate 7. The praying mantis is a carnivorous insect of immeasurable value to gardeners in destroying crop pests.

Plate 8. The red-hot poker plant is a wildflower which has been domesticated.

D

stacked in cones, as in the pine and other conifers ("cone-bearers"). The angiosperms which include the over-whelming number of deciduous or "leaf-falling" species, have many types of covered seeds, popularly called fruits, berries, pods and nuts, although all are properly called fruits in the botanical sense.

When the seeds fall from a forest tree they may land on rocky or overly wet ground where they cannot grow. They need the right amount of moisture and shade to germinate and grow into young trees. Because of these requirements, seeds from a solitary tree on a lawn or in a park have little chance of producing seedlings unless a thick underbrush is kept growing around the base to shield but not choke out the seedlings.

## Deciduous and Evergreen

The greatest natural element with which the temperate forest must contend is drought. The worst and most pro-longed drought does not occur in summer, but in winter, when all of the available water is bound up as solid ice and snow. It is to protect against this winter drought that deciduous trees drop their leaves in autumn. The drop is caused by the flow of a hormonal substance which closes off the microscopic tubes that carry water into the leaf. This loss of required water and minerals causes the leaf to assume its brilliant colors in the autumn. The tree stops manufacturing chlorophyll. With the chlorophyll cut off, other pigments present in the leaves become visible.

**17**

Illus. 13. The secondary grass and shrub growth spreads right up to the edge of the forest.

The leaf is actually a very complex solar panel. It must spread the greatest number of cells to the sun so that the process of photosynthesis may take place. However, since the leaf is so thin and exposes such a large surface area to the drying effects of sun and air, it also wastes the most water of any part of the tree. To prepare itself for the drought of winter, the tree must divest itself of this loss and so drops its leaves in a brilliant shower of color. The long cylindrical needles of pines and firs, both gymnosperms, however, lose much less water and can retain their needles all winter. Such trees are termed "evergreens." However, not all gymnosperms are evergreen—the larch, for example, sheds its leaves—and some angiosperms, such as the holly and the rhododendron, keep their leathery leaves through the winter.

## Succession

The forest has a life cycle, just like any single organism whether plant or animal. At one time in the deep past, all of the now-forested regions of the earth's temperate zone were covered by water. The vast seas receded and slowly the flat plains and hills were washed free of salts (which inhibit plant growth) and after thousands of years the fresh water from the rains dominated the ground water, that is, the water below the soil. The succession of plant communities can still be observed today. Any marsh or pond may sooner or later give way to various stages of plants (successions) and finally become covered over with forest. In the garden, the need to cut grass is not just for the purpose of getting a good-looking lawn—it is the simplest way to retard and prevent the advance of succession. While some of the stages of succession proceed very slowly, others proceed very rapidly and may take place in a single season.

Briefly, the rise of the forest takes place in the following manner. Masses of one-celled algae and long filamentous algae fill up the shore lines of fresh-water ponds and lakes. In time, these form vast mats of green material which die and sink to form thick *detritus* on the bottom. This layer of rich organic material is the basis of all the further stages. The following steps or stages occur with fair regularity in

Illus. 14. Various grasses and annual plants flower and seed in midsummer. They will seed their species for growth the next spring. In the process known as succession, the seeds will often fall on ground that was formerly marshy.

deciduous forest areas, although there may be some local variations from this overall time-table.

1. Emergent plants appear along the boundary between land and water, which is called an *interface*. These have their roots permanently below the water and their green leaves above in the open air.

2. Pond weeds expand into the margins of the pond into the soil held firm by the emergent plants.

3. Marsh plants such as reeds grow out into the shallows until they grow completely across the former pond bottom from shore to shore.

4. Non-aquatic annual herbs and grasses supplant the marsh varieties.

Illus. 15. Yarrow is one of the more common and attractive of North American wildflowers.

Illus. 16. This yarrow has bloomed and seed pods are forming in the tiny white flowers. Yarrow and its unrelated look-alike, the wild carrot, came to North America with the European settlers. Both have run wild over cleared land, roadsides and the edge of the woods.

Illus. 17. Two tall thistles stand dark against the sky on a fall day. The strong stems are covered with sharp thorns. Thistles include several tenacious perennial species that quickly take over open ground and waste land.

5. Perennial herbs, bushes and shrubs appear.

6. The final or *climax* stage is reached and trees grow on the once submerged location.

The climax-stage forest includes not only a stable population of trees, but many lesser plants which take up life among the wet shade of the ground cover.

## Life Cycle of the Forest

Succession is only one aspect of the life cycle of the forest. The other side consists of the perpetual cycle of generation-maturity-death-renewal. This is the cycle familiar in the deciduous forest where new seedlings continually grow among the decaying trunks of old trees

Illus. 18. New shoots grow from a section of willow which had been cut down. The ability of many trees to regrow from cuttings and roots enables forests to renew themselves after forest fires and floods.

knocked down by winter storms and ice. Unlike the deciduous trees, the annual herbs of the forest floor die back each autumn and leave behind their organic materials to enrich the soil and their seeds to carry on the species. The fibrous material of their leaves and stems will be broken down by the process called humification (creation of humus) and their seeds will sprout with the warmth and rainfall of spring.

Some species are so tenacious of life that they will continue to sprout even after they are cut down or a portion of the original trunk remains. The author once cut a section of branch about one metre (3 feet) in length from a pussy-

Illus. 19. A bean seedling sprouts above the soil carrying its original seed coat and large food-carrying cotyledons (first leaves) which produce more food for the young plant by photosynthesis. Although a garden angiosperm, the bean well illustrates the initial spurt of energy which the woodland angiosperms must make if they are to survive.

willow bush. He stripped off all the leaves and used it as a stake in the garden, hammering the bare branch into the ground, top end down. Within a few years a magnificent pussywillow bush had grown from the upside-down stake.

Certain component species of plants in the forest are able to fill *niches* (habitats) offering conditions of seemingly incredible hardship. Foremost of these are the lichens, which are really two plants in one. Each lichen consists of an algae intertwined and living in co-operation with a tough, resistant fungus. These small plants will lie dormant for long periods of drought or heat and sprout and spread when the environment grants them a niche, usually a rock or a tree stump. The mosses are also capable of long dor-

Illus. 20. Tiny plants such as mosses and lichens grow from the humus on the forest floor.

mancy and grow in the groundcover of the forest. Mosses, lichens and ferns often grow in rocky outcrops shaded by forest trees.

## Saprophytes

The last stage of life for any plant is that of dissolution and decay. This process involves a whole host of organisms from bacteria through insects and earthworms, called *reducers*. Among the most important reducer organisms are the saprophytes—plants growing on dead organisms— such as the many varieties of fungi. These plants break down the tough materials which bind the cellulose of the dead tree and release the sugars to be used by them as food.

Illus. 21. Deep channels and tunnels have been carved into an old stump by the action of insects. These activities reduce the tough woody material to digestible forms for bacteria.

The smallest and some of the commonest are the moulds, which combine with the efforts of boring insects to break down the toughest hardwood. Within a few years they can reduce a tall tree to consumable organic materials. The reduced plant material in the form of humus supports the growth of the forest, as well as the garden, and both the annual and perennial plants which grow in the interface between the forest and the garden.

The community of forest fungi is as rich and varied as any other in the plant kingdom. There are many different forms, some nearly microscopic and others with incred-

Illus. 22. A very old cedar stump is surrounded by young trees and shrubs which are slowly taking over the area once drained by the cedar's roots.

Illus. 23. A close-up examination of the cedar stump shows that insects have attacked the softer layers of the dead tree. Holes under such stumps are the primary nesting places of many small animals of the forest.

Illus. 24. The incredible Indian pipe, one of very few flowering plants in North America which have no green material. These young shoots will grow to full height (a few inches) in only 48 hours.

ible shapes and bright colors, which are as large and striking as any flower. One of the most peculiar saprophytes is the Indian pipe, which appears only in areas where some large amount of dead material has been reduced and only after seasons of heavy rain. The Indian pipe is not a fungus, but a degenerate angiosperm that has lost the power to make chlorophyll!

Since the saprophytes have no chlorophyll with which to synthesize foodstuffs, they must dissolve and absorb the food stored in the materials of dead plants or animals which stored such foodstuffs while alive. The saprophytes have no roots but send out into the dead tissue thin filaments of protoplasm called *mycelia*, streams of cellular material which secrete chemicals called *enzymes*. Enzymes break down the complex molecules of the dead plant, releasing the sugars and proteins to be dissolved and absorbed by the saprophyte. The stored-up energy is thus re-used by the saprophytes. It is this same enzymatic action of fungi which is used to make bread dough rise and to turn milk into cheese. By means of other reducer

Illus. 25. The inner layers of the stump have been turned to powder by the action of insects.

**Illus. 26. Wood-boring beetles (middle) prepare to hibernate under the bark of a rotten tree.**

plants, the materials of the climax forest are continually being generated, maturing, dying and being recycled and generated again by long complex associations of inter-active organisms.

# The Garden

A garden is a unique community, a small environment created by man. A garden can only survive with the help of man's continual intervention in ecological evolution. Unlike the forest, which produces and stores great amounts of energy, the garden requires the input of large quantities of energy in order to grow and produce.

A garden is an artificial community, and it is both over-specialized and out of balance. It must have light and warmth, but cannot withstand drought. It must have adequate water but cannot resist flooding. A garden must obtain the proper organic and mineral compounds for growth, but cannot store them for future use. Most of the plants of the forest are perennials that grow year after year, while most of the plants in the garden are annuals, which means that they die off each autumn, leaving their seed behind to perpetuate the species. The garden community really amounts to a temporary seasonal pause in the annual progress of ecological development, and it can easily fall prey to blight, animal predation, insect infestation or human neglect.

The first garden plants were gathered from the wild

Illus. 27. Wild blackberries grow near the woods.

Illus. 28. A field grass going to seed.

and transplanted in the form of seeds to special locations so as to produce a higher density of one plant which could be used for food. The first such domesticated plants were seed grains such as those which can be found growing along the margin of any forest today. In fact, it is the process of humification which results from forest growth that makes gardens possible—that is, the ready availability of humus makes it possible to enrich the garden soil for heavy crop production.

The earliest known gardens were grown by the hill people of ancient Iraq about 10,000 years ago, who grew plants for their edible seeds and berries. Other types of garden plants have been adapted from wild forms to yield food for domestic animals rather than directly for humans. Through the centuries, better-yielding crops have been

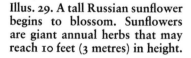
Illus. 29. A tall Russian sunflower begins to blossom. Sunflowers are giant annual herbs that may reach 10 feet (3 metres) in height.

Illus. 30. A ripening sunflower with hundreds of seeds bends toward the ground because its weight may be as much as 5 pounds (2.5 kilos).

Illus. 31. The fully ripe giant sunflower.

Illus. 32. A small pumpkin on its vine swells and turns to a shade of orange.

developed by careful selective breeding. The most productive wild grains produce about 48 seeds for every original seed planted, while the best domesticated species may yield over 2,000 for every one planted! Thus, selected strains produce high yields of seeds.

Garden plants are grown not only for their protein-rich seeds, but also for their fleshy and edible seed coats, such as in the case of the squash and the melon. An extension of the food garden is the flower garden. While some plant varieties such as the money plant are grown for their decorative non-edible seeds and seed pods, others are both edible and decorative, such as asparagus, whose young

Illus. 33. A round watermelon ripens on a bed of straw as parts of the watermelon vine grow right over it.

shoots are a delicacy, while its inedible feathery foliage and bright red fruits are pleasing to the eye. However, it is the flowers with their rainbows of color and sweet scents which are the chief delight of most gardeners. Still other varieties of plants, usually shrubs, are preferred for their full foliage and brightly-colored seeds and seed coats, the firethorn, for example.

## Annuals and Perennials

With the rare exception of asparagus and rhubarb, which are perennials, the overwhelming number of herbs in the garden are annuals. This means that they succumb to the frost in the autumn and have to be replanted again

Illus. 34. Wild garlic ripens into dark red berry-like seed pods.

each spring. They usually begin to decline with the late summer drought and the tension or toughness in their leaves, called *turgor*, lessens, making them all the more susceptible to insect attack.

It is at the time of the summer drought that hordes of insects descend upon them. A good example of this susceptibility is the string bean, which is valuable both as a rich food and also because it is one of the species of plants called *legumes*. The legumes have special sacs on their roots which house a particular species of bacterium. The bacteria in these root chambers are able to *fix* or change free nitrogen from the atmosphere into stable

Illus. 35. The seed pods of the lunaria or "money plant" look like thin silvery coins with the seeds in the very middle.

compounds available as plant food. The bean plant thus adds nitrogen to the soil wherever it grows.

As autumn approaches, the Mexican bean beetle appears in great numbers and starts to reduce the dying bean plants. The beetles eat away the soft parts of the leaves and leave the tougher supporting veins. The fruit is spared, however, for the bean seeds have a wax-like coating which is too hard for the bean beatles to eat. The resulting cut-outs on the leaves are called *fenestrations* from a Latin word for "window." The bean beetle thus aids in recycling the bean plant.

Illus. 36. Asparagus plants produce red, berry-like seed pods which fall to the ground in late summer and develop into new plants the next season.

## Recycling of Animal Matter

In the forest, the organic materials of all living things, including animals, are recycled to be used again. The same processes of reduction and decay that return the molecules of plants to usable form apply to animals. However, in the garden, recycling will not occur naturally because single species are planted in beds and rows to provide high yields of one crop and unwanted species (weeds) are constantly removed, or destroyed along with animal pests. This process of selection in the garden eliminates some of the species required for natural organic recycling.

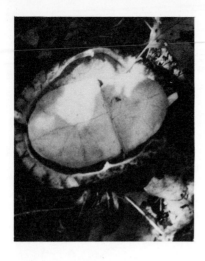

Illus. 37. The hollow and bleached shell is all that remains of a box turtle which died of old age in the forest.

Just as the succession of the forest over other stages of plants comes in an orderly progression and the decay and reduction of plants requires many associated organisms, so the reduction of animals takes many participants. A long procession of bacteria, fungi and insects feast upon dead animals. The first step after the animal dies occurs when the internal bacteria begin to make chemical changes in the tissues. Within a few hours, if the organism is visible from the air, scavenging birds, such as crows and vultures, will descend to devour what they can. What is left by the birds will continue to decay from bacterial action.

The sour odor of the bacterial wastes and chemical changes will attract flies, which will lay their eggs in the dead body. The eggs soon hatch into small segmented white worms called *maggots*, which live on the decaying flesh. The maggots are often joined by ants and carnivorous beetles. Some species of beetles actually dig out

Illus. 38. Fluffy molds grow on a dead branch lying on the forest floor.

under and around the carcass and bury the dead animal. When the organism is buried, the beetles and their *grubs*, or larval forms, devour the remaining soft parts. Over a period of time all but the calcified bones are consumed. Since the decay process releases dissolved organic substances into the soil around the carcass, species of saprophytes will often appear from the humus where an animal body has decayed.

## Fertilizers

Gardeners must increase the natural growth and sustenance of garden plants without waiting for the slow processes of animal decay to enrich the soil naturally. Instead, they often utilize processed animal products, such as

**41**

Illus. 39. Dead organic materials under the soil support the growth of these small fungi.

manure, blood or oils from cows, sheep, birds and even fish. These animal components supply the needs of garden plants and aid the growth of domestic vegetables and flowers. However, much of this material does not get incorporated into the humus because the natural reducers are either very few in number or there are none at all. So, much of the overly rich animal material goes off as a pollutant in the rain-water that washes the garden, and ultimately aids the growth of unwanted algae in the streams and rivers through which it flows, before entering what are called in ecology the final receiving waters, which may be a pond, lake or ocean.

42

Illus. 40. A puff-ball fungus grows through a hole in a fallen oak leaf and its root-like mycelia feed on the pile of dead leaves below.

Illus. 41. The shoots of another type of fungus sprout like fingers from a dead tree.

Illus. 42. A bright yellow bracket fungus grows on a dead hickory branch.

Illus. 43. The underside of the bracket fungus is covered with tiny dot-like spore-producing bodies. These spores are much smaller than the seeds of green plants and are spread far and wide by the wind.

Illus. 44. Very large white cup-shaped fungi grow on the woody layers of a high stump.

Illus. 45. A common mushroom or toadstool form of fungus, this large specimen sprouted out of a cut lawn in one night in mid-summer after a period of rain. It is growing out of a piece of rotted root just below the surface of the ground.

**45**

Illus. 46. The darker recesses of the holes bored by insects in a stump show several varieties of fungi beginning to spread through the wood.

# The Forest-Garden Interface

The marginal area where the forest and the garden overlap is a very special and fascinating environmental zone. It is usually marked by a plant community which includes tall grasses, shrubs and young trees. This is the preferred nesting place of all types of small animals, since it provides them with a ready food supply and does not require them to travel far from the shelter of their woodland burrows where they are safe from predators or other dangers.

## Dwellers in the Interface

Many types of forest creatures can be found in the forest-garden interface, ranging from the tiny shrew to the white-tailed deer. Most temperate forests have game trails and narrow pathways through the interface area. Strange as it may seem, most forest animals range over rather small areas. The author has watched a mature female white-tailed deer raise her twin fawns every season for a decade in a small patch of hedgerow and trees in an interface area. This deer does not range more than a half mile from its bedding area throughout the year.

Illus. 47. A female robin cocks her head as she "listens" for worms below the ground surface.

A rabbit or woodchuck (marmot) will often feed in the same pasture or garden within less than 100 feet (32 metres) from its burrow at any time. A box turtle will remain in an even more restricted territory throughout its lifetime. The location of berry bushes or some other concentrated food supply will occupy a turtle for most of the summer within a 15 square yard area (12.5 metres²). Snails, slugs and spiders have even more limited areas within the interface.

Most types of songbirds nest in the interface areas and feed upon the insects and seeds available in the garden, since they are dependent on tree branches for protection of their nests and young. The ground cover of the forest and interface is especially important to the survival of

**48**

Illus. 48. A pine cone is typical of the seed pods of conifers. The individual spruce and pine seeds grow up under the overlapping flaps of the cone. These flaps dry and spread in the autumn, dropping the seeds to the ground. In time the seeds will sprout and saplings will grow up to provide shelter for birds and insects.

reptiles and amphibians, which are unable to control their body temperatures, as birds and mammals can. During periods of drought or in those seasons of the year when morning chills may develop, these "cold-blooded" animals must return to the ground cover. Toads are especially susceptible to dessication, being dried out, and for food must seek flying and crawling insects in defined territories in the interface.

On the trunks and within the bark of interface trees, many flying insects make their final moult to become airborne. Many individual insects die in the open grass or garden because they cannot find a suitable perch above the ground where their newly unfolding wings can dry out and from which they can launch themselves into the air.

Illus. 49. A branch of a heavily laden firethorn bush hangs in the chill air of late September.

## Interface Flora

Many types of plants and their younger forms do best in the interface environment. As the young tree seedlings reach a height of one foot (30 cm.) they begin to look like miniatures of their species. At a height of 8 to 10 feet (roughly 3 metres) the young tree, now called a sapling, will begin to dominate an area and send out third (tertiary) and fourth (quaternary) branches from the primary and secondary branches which it has already grown. In the autumn it will drop its leaves and appear as a tall whip-like trunk with a few bare wisps of twigs for branches.

Illus. 50. The two stages of a bean seed pod. The upper bean is beginning to break open; note the holes eaten by bugs. The lower pod has already begun to dry and burst, and the ripe seeds will drop to the ground.

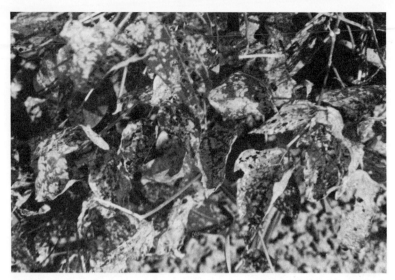

Illus. 51. Bean plants are a basic and vital source of food for a whole community of animals and insects in the garden. These bean leaves are at the end of their season and the autumn drought has left them weakened and susceptible to extensive insect damage. The bean beetles have removed all of the fleshy parts of the leaves and left the tougher veins.

Such saplings are very susceptible to damage and permanent bending by ice storms. While the saplings in the interface have the best chance of survival because of lack of competition for sunlight and minerals, those on an interface open to the north may be so severely damaged by winter storms that they will succumb to insect and fungi attacks in the next growing season.

Since many garden plants are originally from an interface environment or require shading by trees, a zone similar to the interface is often planned and maintained by man. In such an interface, whether natural or man-made,

52

Illus. 52. The herbs and grasses found on the southern edge of the forest are noticeably different from the plants of the northern interface.

there is a decided difference in the communities of plants and animals which grow on the northern side and those which grow on the southern side of the forest. For example, there are many plants which, although sun-loving, require the rich humus of the forest floor. These are found in the interface, usually on the southern side, and in the small occasional forest clearings. Some of the choicest wild-flowers and herbs grow best under these special conditions.

## Nature of the Interface

The interface is the most dynamic environment of the forest and the garden, for it offers more sunlight and water than the forest and more shelter than the open ground of the garden. It is also the most competitive environment

**Illus. 53.** Hoping to escape notice, a young robin "freezes" among the fallen leaves and low branches of the interface.

where the successional stages are changing constantly. As the forest expands, new territories for forest animals and plants are provided. It is among these partially established territories that the predators have their best opportunity. A live trap set at the interface will always achieve the best results, capturing the greatest variety of small animals.

The interface is a truly unique and fascinating environment, but it is one which can be destroyed very easily if not properly tended. It must be remembered that the precise layering of forest mulch in all of its strata requires a number of seasons. Burning, raking or disturbing this humus can destroy the delicate balance of humification organisms. The same is true with stripping back the ground cover or wantonly cutting the saplings in the interface between forest and garden.

# Some Animal Communities

Many, many hours of enjoyment can be derived from watching the actions of the animals of the forest and garden. The smallest and perhaps the most important are the soil nematodes, small segmented worms less than 1/5 of an inch (0.5 cm.) in length. They are among the most common animals on earth—it has often been said that if every animal and plant in the world were to disappear and leave behind their nematode parasites, each organism and each species would be visible and identifiable from its nematodes. Nematodes are basic to the breakdown and recycling of organic materials in the soil. Of nearly the same size and lesser frequency are the soil mites, tiny spider-like arthropods that are found in nearly every clump of forest floor no matter what the environmental conditions. These tiny creatures, only 1/25 of an inch (1 mm.) in length, devour great masses of bacteria, insect eggs and loose remnants of biotic substances from the decay of plants and insects. They are the primary scavengers of the forest humus.

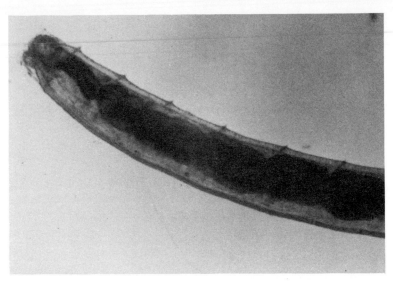

Illus. 54. A nematode worm seen under low power magnification. The nematodes are one of the most common organisms on earth — 1,000 or more may inhabit one square foot (940 cm²).

## Earthworms

But the most important of all the soil creatures is the earthworm, which carries on two activities which are vital to the continuation of the humification cycle and the continued fertility of the soil. First, the earthworm brings down loose particles of organic materials from the surface of the ground by night and devours them in its burrows. Secondly, the earthworm grinds and ploughs its way through the soil, where it leaves its own body secretions and brings up masses of bacteria and digested substances to the surface.

Illus. 55. An earthworm plows its way through a burrow seeking food.

Illus. 56. An earthworm casting.

Earthworms are found in greatest abundance in undisturbed forest and interface humus and soil. Disturbing the garden soil kills both adult worms and their eggs and destroys their burrows. The greatest possible worm population is needed in a garden mulch pile or compost heap in order to keep the humification cycle intact. This humification cycle will provide continual organic matter in a reduced form for fertilizing the garden, and the key factor of the cycle is the earthworm.

## Molluscs

There are many other invertebrate animals in both the forest and the garden and among the most interesting of them are the molluscs, particularly the land snails. Snails

Illus. 57. A land snail makes its slow way across a leaf growing close to the ground.

Illus. 58. The head of the hornworm has a deceptive eyespot which makes the harmless vegetarian look more fearsome to predators.

eat plants such as algae which they scrape from rocks and soil with a rasp-like organ. They have fully developed circulatory systems and nervous systems and are hermaphrodites, that is, each individual snail has both male and female sex organs. However, no snail fertilizes its own eggs, which are laid in groups of several dozen and hatch in three to four weeks. The young have very thin shells and must survive the first winter's hibernation. The age of snails can be determined by the growth rings on the shell.

## Insects

For sheer volume and number of species and individuals, the most obvious organisms are the insects. They are always busy chewing their way through tons of green plants in summer and the evidence of their depredations is everywhere. Few observers realize, however, that the

Illus. 59. A moth larva is nearly lost in the grass because of its camouflage.

Illus. 60. A beetle larva feeds on the needle-like leaves of an herb.

Illus. 61. Bean beetle larva (white spot in middle) is passing through the last stage before adulthood.

worst enemy of an insect is usually another insect. For example, the long green hornworm which attacks tomato plants in the garden is preyed upon by the paper wasps.

There are a number of different life cycles which insects follow. Since every insect species must shed its tough outer skin at least once in its life cycle as it grows, each of these sheddings or "moults" marks a different stage in its life cycle. Some insects change their whole body form during their life cycle and start as a caterpillar which later is altered in form to that of a flying insect such as a butterfly or moth. This total alteration is called *metamorphosis*. When an insect passes from an egg to a larva to a pupa (the resting stage before flying adulthood) to an adult, it

**61**

Illus. 62. The large boring larva of a moth hangs beneath a twig as it begins to spin its cocoon.

Illus. 63. The all-black woolly bear caterpillar was once thought to be a sign of an especially hard winter, but it is simply a color phase of one variety.

Illus. 64. A young grasshopper perches on a flower stem. The grasshopper goes through a series of moults, each stage resembling the final or imago form but smaller and of reduced proportions.

undergoes a complete metamorphosis. This is commonplace with many caterpillars.

When an insect passes from the egg through a form called a nymph to the adult form it has a *gradual* metamorphosis, as is the case with many dragonflies, which develop from waterborne or submerged nymphs. When the insect passes from egg to a series of one or more gradually larger but identical forms or stages, it has no metamorphosis. Such insects go through many sheddings or moults, the stages between each moult being known as *instars*. The grasshoppers and walking-sticks pass through many instars but have no metamorphosis.

Illus. 65. A walking stick makes its way up a garden vine.

Among the largest group of insects are the bugs and beetles, with their hard coats and large external feeding apparatus. A fine specimen is the solitary stag beetle with its enormous pincers used to fight off rivals for its territory. The territory is usually a rotten stump where the beetle takes several years to grow from egg through many instars to adulthood. Included in this group are such pests as the Japanese beetle, one of the most destructive species of

Illus. 66. A fearsome-looking beetle of rotten tree stumps is actually
harmless.

Illus. 67. Japanese beetles, one on
top of the other, can quickly
destroy a rose in the garden.

Illus. 68. A female beetle lays eggs on a plastic greenhouse sheet.

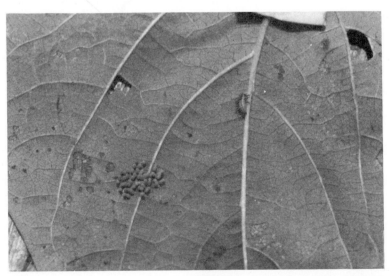

Illus. 69. The eggs and larva of the Mexican bean beetle on the underside of a bean leaf.

Illus. 70. The eggs of a true bug. These eggs will hatch into a young bug precisely like the adult but smaller, and the organism will not pass through an intermediary stage.

Illus. 71. A leaf-eating beetle works its way along the margin of a leafy vegetable.

the garden and the bane of most domestic plants. There are many other species of crop-destroying weevils and beetles, but there are also some that are not harmful. Of unique importance are the different species of ladybirds which prey upon other crop destroyers. It has been estimated that in its lifetime one such insect will devour over 3,000 plant parasites, such as scale insects.

A number of wasps lay their eggs inside the tissues of leaves or stems where they set up a chemical irritation which causes the plant cells to grow outward into a large round chamber. These chambers are called galls and the wasps are called gall wasps. The eggs of insects are often very hard to identify, but this is not the case with the gall wasp.

Illus. 73. Galls caused by
the irritation of wasp
larvae in the tissues of the
leaf.

Illus. 74. The gall has been opened to reveal the wasp larva inside.

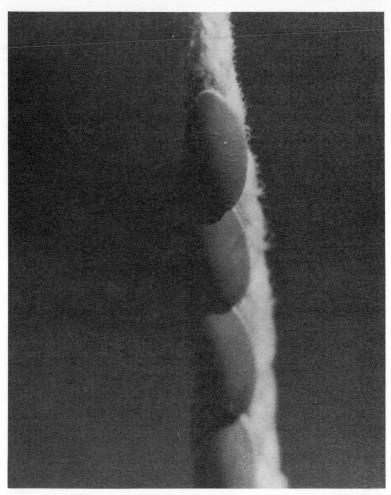

Illus. 75. The eggs of a leaf-eating insect are attached to a stalk. Many insects winter over in the form of the egg of the next generation. Thus, while the individuals die, the species is maintained in its most energy-conserving and resistant form as eggs.

Illus. 76. An aphid lion (left), the larva of the lacewing, seeks and destroys aphids (dark clusters at right).

Illus. 77. A tiny meadow butterfly (middle of photograph) blends in with its surroundings so as to be nearly invisible.

Many insects are so modified in form that they look like part of their environment, resembling leaves or twigs, and so avoid detection by birds and other predators. As the late summer drought approaches, many insects go into their cocoons to pupate and complete their metamorphosis. These may be nothing more than rolled-up leaves cemented with secretions or they may be elaborate woven containers in which the sleeping insect will pass the winter.

While most insects are mere plant pests in the garden, some few are harmful to human beings. One of these is the mosquito, which may be a carrier or *vector* of human disease. The gardener and the student of the woods should

Illus. 78. Bald-faced hornets work on building a papier-mâché nest before winter sets in, when they will cluster inside and pass the cold months in a sort of stupor.

Illus. 79. A worker ant removes larvae to a safe place after its nest has been disturbed.

Illus. 80. A giant bumblebee extracts nectar from flowers.

observe closely so as to learn to distinguish insect friends and foes.

Of all the insects, the social species which live and work together as one remarkably well-ordered unit are always the most fascinating. The bees not only manufacture honey from the nectar of flowers but also help in the pollination of plants. Some of the more common bees of the interface, such as the large hairy bumblebee, nest in the ground, while the tiny species commonly called "sweat bees" spend their lives in or near ponds or streams in the forest. Ants scavenge on the dead organisms on the surface of the ground while others actually keep flocks of small insects called aphids like cows for their sweet secretions.

Illus. 81. An exceptionally large nursery web spider sits astride the silk lines which hold its eggs and young.

## Arachnids

The most industrious predators upon insects are the arachnids, or spiders and their relatives, which are different from all insects in that they have 8 legs (insects typically have 6) and live in solitary territories, not in communities. Two types of spiders are widely distributed in gardens, the crab spiders, which hold their legs out to the sides like those of marine crabs, and the large nursery web spiders. The webs of spiders have interested human beings for thousands of years—they figure in many myths and fables. A woodland species, the spined microthenas, with their strange irregularly-shaped bodies, are common in the forest where they construct large webs to trap flying insects. The large females of this species are constantly re-

Illus. 82. A magnificent and intricate spider web is highlighted by cold morning dew which makes all of its strands visible.

Illus. 83. A red spider works on its strong, silken web, strung low to the ground.

building and repairing their webs and usually have a brood of nearly microscopic young hidden nearby. The micrathenas will build nests in large trees—the author has collected them a full 10 feet (3 metres) above the forest floor.

Illus. 84. The two layers of a sheet web cling to a pine tree.

Illus. 85. An enlarged close-up photo shows the detail of the anchor lines which support the finer cross-lines of the sheet web.

The orb web–building acacesia is a reddish-brown species of the forest and low bushes of the interface.

Less common but equally interesting are the sheet-web weavers, such as the frontinellas, which make a horizontal bag-like web often in pines or other trees with stiff leaves. Similar to this horizontal web is the funnel web built low to the forest floor by the cicurina group of spiders. These spiders await any intruder and bolt forward to seize and capture their prey. Like all spiders, they inject enzymes into their victims and suck out the liquefied biotic fluids. Some spiders build only a shelter web in the higher branches of bushes and saplings in the interface zone. This

Illus. 86. A funnel web leads into a silken tube lower in the shrub or grass where the spider lies in wait for insect victims.

is the particular environment of the pisaurinas, which capture their prey by swift leaps.

Often walkers in the forest will see the strange awkward harvestmen or "daddy longlegs" spiders of the family Phalagiidae running up the bark of trees or over the dead leaves. These are scavengers and are often the first creatures on the scene whenever an animal is killed. They are solitary and are incredible in their skill at moving about with their bodies underslung beneath their long legs.

The author has often fed these spiders tiny bits of meat and any other animal substances available. They will travel regular foraging trails in search of food. It is actually

Illus. 87. A large garden spider adds new strands of secreted silk to its web.

Illus. 88. A yellow crab spider walks crab-like across a leaf in search of small leafhoppers.

Illus. 89. A large microthena spider repairs her web.

Illus. 90. A small garden spider has encased its eggs in a web suspended from the underside of a leaf.

Illus. 91. A young harvestman feeds on the animal fat adhering to the outside of a French-fried potato. These beneficial arachnids will return to a feeding place almost like a household pet.

possible to locate an individual harvestman and get it to return to the same spot to feed many times over. Although the overwhelming number of spiders are harmless, one related group of arachnids, the mites, is quite harmful. These are widespread in the woods of North America in July and August, as are the related ticks. These arachnids

Illus. 92. A large harvestman or "daddy longlegs" makes its way across fallen leaves.

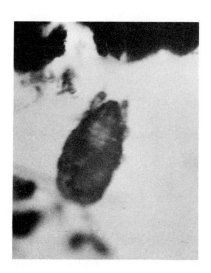

Illus. 93. A microscopic soil mite, one of the many arachnids which help to break down plants into new humus. These tiny creatures are plentiful but rarely seen.

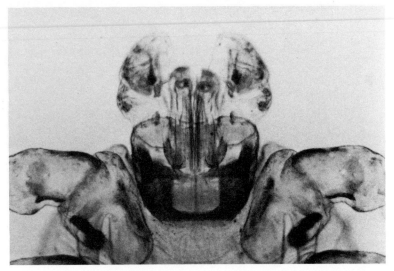

Illus. 94. The mouth parts of the common tick. This magnified photograph shows the powerful jaws with which the tick fastens to its prey and through which it draws blood.

have strong sucking mouth parts and can work their way almost totally under the skin to gorge on blood and lay eggs.

## Vertebrates

The reptiles and amphibians of the forest and the garden are most active at night. They eat massive volumes of insects and stay hidden most of the year in the interface zone. The common box turtle is a good subject for study,

Illus. 95. A mother box turtle watches over a hatchling.

Illus. 96. An old male box turtle feeds on wild strawberries, one of the species' most esteemed foodstuffs.

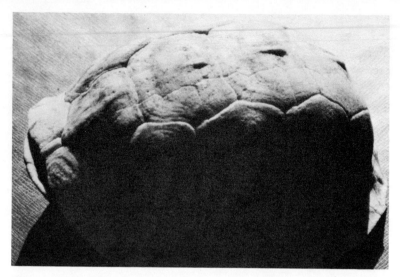

Illus. 97. The shell of a dead male box turtle displays the box-like shape from which the animal gets its name. Note the numerous ridges, which support and strengthen the shell.

as it rarely goes very far and will return to the same area year after year if left alone. There are few garden animals which give as much pleasure or are as interesting to feed and observe as the box turtle. However, these animals should be restrained from approaching strawberry or tomato plants or other thin-skinned garden crops, since these are high on the box turtle's list of preferred foods. Box turtles should not be brought indoors in winter, but should be allowed to dig down in the forest or interface to hibernate.

Illus. 98. The moment of hatching of a box turtle.

Illus. 99. A box turtle only a few weeks old crawls through the grass near the edge of the garden.

Illus. 100. The greatest predator of the garden is the garter or garden snake.

The garden snake, like the box turtle, will stay within a small territory for many years, especially if a small pond or watering place is provided near the interface. Garden snakes are totally harmless and eat mostly earthworms, insects and toads. However, the age-old human fear of snakes is still strong and many suburban areas have been completely cleared of snakes. Although some garden snakes will withstand captivity and eat well, they should be released by late summer to fill up on the food they will need to store in their bodies for hibernation. They will often hibernate in the foundation walls of outbuildings or in mouseholes near houses in order to take advantage of the heat coming from the building.

Akin to the reptiles but much more numerous and welcome are the birds. Their variety changes from locality to

Illus. 101. A little girl is able to sneak up next to a young robin, which has the instinct to avoid detection by standing still so as not to be seen.

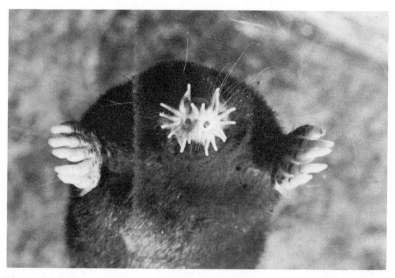

Illus. 102. The star-nosed mole with its two most important adaptations, its large digging fore-paws and weirdly expanded but super-sensitive nose.

Illus. 103. The blurred object is the mole making its way back underground at a furious pace. This clumsy-looking creature can move very fast indeed, if alarmed.

locality, but the American robin, the arch-enemy of earthworms, is found through the United States and Canada. Just under the ground surface of many lawns and gardens live the moles. These small but powerful mammals with their instinct for digging are rarely seen unless dug out of the ground or forced out of their burrows by rainwater. Their forelimbs and noses are splendid adaptations to their subterranean life cycle and they make pests of themselves by tunneling under lawns and gardens, causing unsightly mounds and ridges.

The most widespread of the mammals and the largest plant predator of the garden is the rabbit. So fond are

Illus. 104. The skeleton of a rabbit which died in the forest has been stripped down to the bones by ants, flies and beetles.

rabbits of such plants as beans and lettuce that they will often nest and bear their young among the rows of vegetables. They will nest over the winter in deep burrows in the forest or interface and raise vast numbers of young. Since men have destroyed most of the original wild predators, such as the fox and the wolf, which preyed on rabbits over much of North America, there is a vast population boom among rabbits. Although shooting them may not be to everyone's taste, they should be hunted in the autumn to reduce their numbers and keep the strain healthy.

The most successful of all of the creatures of our par-

Illus. 105. A rabbit burrow in a vegetable garden—the naked, blind young can be seen in the upper middle.

Illus. 106. A child holds three new-born rabbits.

Illus. 107. The bark is falling off an old rotted stump as a frightened deermouse (middle) dashes for cover.

ticular forest and garden is the deermouse, or American deer-footed mouse. These quick little creatures will make their homes almost anywhere from hollow stumps to the eaves of houses and barns. They will eat almost anything and are the best adapted of all of the animals of forest and garden to the coming of human settlement. The stories and tales of deermice and their exploits in search of food are numerous and more often than not, perfectly true.

Not all of the animals to be observed in the interface or the garden have been mentioned here. Deer, raccoons, squirrels, opossums and others that occasionally appear in these areas are woodland animals that have developed a taste for man's produce or are driven by a scarcity of their natural food supply to invade the premises of human beings.

# Conclusion

Before the coming of the European settlers, most of eastern North America was covered by a mantle of virgin forest, only here and there broken by open areas, either natural or sparingly cleared by American Indians to grow small crops of maize, beans and squash.

By the 19th century the interface zones between forest and cultivated land had grown enormously in total area. With the opening up of the rich agricultural lands to the west, much cleared land in the East was later allowed to return to second-growth forest, or was turned into suburban or exurban residential areas. The result was that, in many parts, the once receding forests began to spread again and to come into ever greater contact with gardens.

Today, more and more former urban dwellers are living on the fascinating edge of the interface. It is to be hoped that this book will enable them to understand better the forces, some visible, others not, that are at work along the boundary between the mighty forest and the fragile garden.

# INDEX

Mexican bean beetle, 38, 66
micrathena spider, 81
  spined, 75, 77
mites, 82–84
  soil, 55, 83
mole, 89, 90
  star-nosed, 89
molluscs, 58–59
money plant, 35, 38
mosquito, 72
moss clump, A
mosses, 24–25
moth larva, 60, 62
moulds, 26
  fluffy, 41
moults, 61
mouse, deer-footed, 93
mushroom, 45
mycelia, 29, 43
"naked seeds," 16
nature of the interface, 53–54
nematodes, soil, 55, 56
niches, 24
nursery web spider, 75
perennials, 36–38
Phalagiidae, 79
photosynthesis, 14–15
pine cone, 49
pisaurinas, 79
plants
  bean, 52
  emergent, 20
  marsh, 20
plants and animals, 9–13
plant succession, 7
ploughing action, 6
pods, seed, 37, 39, 49
poker plant, red-hot, D
pond weeds, 20
praying mantis, D
protoplasm, 29
puff-ball fungus, 43

pumpkin, 35
Queen Anne's lace, 21
rabbit, 48, 90–92
recycling of animal material
  39–41
red-hot poker plant, D
red spider, 77
reducers, 25–30
reptiles, 9, 49, 84–88
robin, 54, 89, 90
  female, 48
  hatchling, C
Russian sunflower, 33
salt, 19
saplings, 52
saprophytes, 25–30
secondary grass, 18
secondary growth of trees,
  16
seed, 16
  bean, 51
  coat, 11
  germination, 17
  pods, 37, 39, 49
seedling, bean, 24
sheet web, 77, 78
shrubs, 36
  growth, 18
snails, land, 58–59
snake
  garden, 88
  garter, 88
soil mites, 55, 83
soil nematodes, 55, 56
solitary stag beetle, 64
songbirds, 48
spiders, 75–84
  crab, 75, 80, C
  garden, 80, 81
  micrathena, 81
  nursery web, 75
  red, 77
  spined micrathena, 75, 77
  web, 76

spores, 44
star-nosed mole, 89
string bean, 37
stump, 26, 29
  cedar, 27
succession, 19–22
  plant, 7
sunflower, 33, 34
  giant, 34
  Russian, 33
sweat bees, 74
thistles, 22
ticks, 82–84
toadstool, 45
trail, game, 12
trees
  branches, 15
  deciduous, 16, 17–19
  evergreen, 17–19
  secondary growth, 16
  tall, 4
turgor, 37
turtle, box, 40, 48, 84–87
undergrowth, 6
vector, 72
vertebrates, 84–93
walking stick, 64
watermelon, 36
web, spider
  funnel, 79
  sheet, 77, 78
web spider, 75
weeds, pond, 20
wild blackberries, 32
wild carrot, 21
wildflower, 21, D
wild garlic, 37
wild Indian ginger, B
willow, 23
wood-boring beetles, 30
woodchuck, 48
worker ant, 73
yarrow, 21